Cancer Patient's Guide to Radiation Therapy

RICHARD STEEVES, M.D., PH.D.

MEDICAL PHYSICS PUBLISHING
Madison, Wisconsin

Copyright 1992 © by Richard Steeves

All rights reserved. Reproduction in any way by any means of any part of this book without permission of the publisher is prohibited by law.

Published by:
Medical Physics Publishing
732 N. Midvale Boulevard
Madison, WI 53705
608-262-4021

ISBN: 0-944838-26-X

Library of Congress Cataloging-in-Publication Data

Steeves, Richard A.
    A cancer patient's guide to radiation therapy / Richard Steeves
    p. cm.
    Includes index
    ISBN)-944838-26-X (pbk.): $9.50
    1. Cancer--Radiotherapy--Popular works.   I. Title.
RC271.R3S743   1992
616.99'40642--dc20   92-1624
        CIP

Foreword

One out of approximately 500 people in the United States are diagnosed with cancer each year. Although the incidence of cancer is growing, our knowledge of how to treat cancer with surgery, chemotherapy and radiation therapy, alone or in combination, is also growing. Less than half of all people who learn they have cancer this year will die from their cancer. As a cancer patient, this book will help you understand cancer and how to cope with the disease.

Modern radiation therapy helps cure a significant number of these patients. This book will tell you how to choose a radiation therapy center, if this mode of treatment is recommended. Dr. Steeves explains the techniques and instruments that have permitted radiation therapy centers to cure more cancer patients each year. He speaks with authority as a recognized leader in the field of radiation oncology.

Paul P. Carbone, M.D., D.Sc. (Hon.), F.A.C.P.
Professor,
Departments of Human Oncology & Medicine
Director, UW Comprehensive Cancer Center
May 1992

Preface

If you think you have cancer, you could be having a lonely and terrifying experience. Panic sets in when you realize that you have no ready source of information about cancer or its treatment. This book will help you understand what is happening to you. Focus your attention and gain a little knowledge now, so that you can be actively involved in your care, become an active partner with your physician, and bypass many of the uncertainties that would result if you were a passive patient.

In my daily contact with cancer patients I am often struck by their mental suffering, mostly from fear of not knowing what is going to happen to them. Most patients are unable to control this fear because they are not well-informed about their disease and its treatment.

If you are a cancer patient, a close friend or a relative of a cancer patient, or if you are worried that you might have cancer, then this book is for you. Here you will find answers not found in most books about cancer. I hope it will reduce your fear and anxiety.

As a result of the highly publicized accidents at Three Mile Island and at Chernobyl many patients worry about the potential dangers of x-rays. Therefore, I emphasize how safe radiation therapy really is and how well tolerated it is by most cancer patients. If I manage to prevent unnecessary fears from lowering the quality of life of just a few cancer patients, I will have fulfilled my goal in writing this book.

> Richard A. Steeves, M.D., Ph.D.
> Madison, Wisconsin
> May 1992

Acknowledgements

I am especially grateful to Dr. John Cameron and Julie Bogle for their encouragement and helpful suggestions during the preparation of the manuscript.

Thank you also to Siamak Shahabi, Ph.D., Deborah Tinker, Maria Butkus, and Gladys Meier for their numerous helpful suggestions. A special thank you also to Doris Blancke, Eileen Healy, Alan Bogle, Melodie McPherson, Jennifer Weaver and Dr. Paul Harari for allowing us to photograph them.

Considerable assistance was also provided by Lori Rosin and Diane Chechik of Sarasota, Florida, Dr. Judith Stitt, Kris Saeger, Mary Burkhammer, Dr. Paul Carbone and by my wife Eliane.

Contents

CHAPTER 1
What Is Cancer? What Are Some Typical Symptoms? 1
    Cancer Cells Don't Rest
    Cancer's Warning Signals Spell C.A.U.T.I.O.N
    Early Detection is Important
    Cancer Spreads to Other Organs

CHAPTER 2
Coping with Cancer 5
    The Shock
    Reactions
    Finding Hope
    Developing a Plan
    What is a "Cure?"
    Grief
    How Do You See Yourself?
    The Role of Family and Friends

CHAPTER 3
Whom Should You See for Advice and Treatment? 12
    Doctors are Not Equal
    Sources of Information
    Getting Another Opinion
    Testing and Waiting
    Talking to Your Doctor

CHAPTER 4
Where Should You Go for Treatment? 16
    University Hospitals
    Comprehensive Cancer Centers
    Major Medical Centers

CHAPTER 5
How Is Cancer Usually Treated? 20
    Surgery
    Radiation Therapy
    Chemotherapy

CHAPTER 6
Radiation Therapy for Treating Cancer..................28
  The Role of the Radiation Oncologist
  Some Questions You Might Ask
  Planning Your Treatment
  Getting Simulated
  Your First Treatment
  In the Treatment Room

CHAPTER 7
What Are the Side-Effects of Radiation Therapy?......45
  Side Effects During Treatment
  Post-Treatment (Late) Side Effects
  Radiation-Induced Cancer

CHAPTER 8
Special Types of Radiation Therapy..................50
  Brachytherapy
  Stereotactic Therapy
  Intra-Operative Radiation Therapy

CHAPTER 9
Other Types of Therapy..............................54
  Hyperthermia
  Immunotherapy

CHAPTER 10
Commonly Asked Questions About Radiation Therapy....60

CHAPTER 11
Where to Find Additional Information................67

GLOSSARY..71

APPENDICES
Comprehensive Cancer Centers.......................82
Radiation Oncology Residency Training Hospitals....86

INDEX..95

CHAPTER 1

What Is Cancer? What Are Some Typical Symptoms?

Cancer simply means malignant or uncontrolled cell growth of any tissue in the body. It excludes benign or limited-growth tumors such as warts or moles. There are many types of cancers but they can be divided into the following four categories:

- **carcinomas**: cancers that arise in the skin or in the lining of the digestive, breathing or urinary systems, or in any gland, such as the breast, prostate, pancreas, testis, or ovary
- **sarcomas**: cancers that begin in the body's frame, such as bone, cartilage, muscle, fat, or fibrous tissue
- **lymphomas**: cancers that arise in lymph nodes
- **leukemias**: cancers that begin in the bone marrow

The word cancer commands fearful respect, partly because it can be difficult to cure. With recent improvements in diagnosis and treatment, however, more than half of all cancer patients have no evidence of disease five years after treatment. The vast majority of such patients are, in fact, cured.

Many doctors, to avoid alarming their patients, use the word *tumor* instead of cancer. A tumor can be either benign or malignant, so if your doctor's use of this term is confusing to you, ask for clarification.

Cancer Cells Don't Rest

Cancer starts when a single cell loses its ability to be controlled by the normal regulatory signals around it. After cell division the two "daughter" cells do not go into a resting period as normal cells do. Instead, they immediately start dividing. The time it takes for cancer cells to divide varies quite a bit, from a few days to a few months. It might take several years for a clump of cancer cells to grow big enough to be seen or felt. After about 33 divisions the tumor will measure an inch or two in diameter. At this point the patient may notice symptoms such as a lump or difficulty with breathing, swallowing or urinating. Some cancers grow into the skin and become a bleeding sore. Be alert for the following symptoms, also listed by the American Cancer Society:

Cancer's Warning Signals Spell C.A.U.T.I.O.N.

- Changes in bowel or bladder habits
- A sore that does not heal
- Unusual bleeding or discharge
- Thickening or lump in the breast or elsewhere
- Indigestion or difficulty in swallowing
- Obvious changes in a wart or mole
- Nagging cough or hoarseness that will not go away

Early Detection Is Important

The easiest way to detect cancer early is to be familiar with your own body. Every time you bathe you should run your hands over your entire body, including your neck, armpits and groin, and carefully note any lumps or sores. If you're a man you should feel your testicles; if you're a woman you should palpate your breasts. If you do this regularly, you will be aware of and sensitive to any changes in your body. See your doctor immediately if you have any doubts or concerns.

Another simple way to detect cancer, especially in the urinary system or colon, is to purchase a kit at your pharmacy. The kit will detect hidden blood in your urine or stool. Because of the frequency of breast cancer, women over 50 years of age should have annual mammograms. If there is a family history of breast cancer, it is wise to begin annual mammography at 35 to 40 years of age. Men over 50 years old should have tests for prostate cancer annually. Heavy smokers, men or women, should have a chest x-ray each year.

Cancer Spreads to Other Organs

If cancers were dangerous only because of their unregulated growth, then we might expect that a cancer could be removed by a surgeon and that would be the end of it. Unfortunately, many cancer cells have a second undesirable property: they travel and set up housekeeping somewhere else.

Cancer's ability to spread and grow in new places, called *metastases*, occurs via two pathways: the blood circulation and the *lymphatic system*. (See the Glossary for a more complete definition.) Most people know that cancer cells can travel through the bloodstream to almost any part of the body. Many do not know that cancer cells can also travel a short distance to lymph nodes through the lymphatic channels. Lymph nodes are located just under the skin in the neck, armpits, and groin, and more deeply behind the breast bone or sternum, and on either side of the spine. Some cancers, such as sarcomas and leukemias, have a tendency to spread mainly through the blood stream. Other cancers, such as carcinomas and lymphomas, often spread through the lymphatic system. There are, however, many exceptions to this generalization. Your doctor may need to consider both pathways in your situation.

CHAPTER 2

Coping with Cancer

The Shock

If you or anyone in your immediate family develops cancer, the whole world may seem to crash suddenly at your feet. Most people are accustomed to a predictable way of life. Suddenly schedules go up in smoke. Relationships may become strained. Worries about bills and money are often a source of turmoil. The biggest problem of all is the uncertainty—what will be the outcome of the disease? When will you know?

Reactions

Of course, the person with cancer is affected most of all. Denial may set in, then depression, followed by anger. A typical reaction to cancer is, "Why did this happen to me?" Resentment is another common reaction, directed perhaps toward family members and friends who don't have cancer, or toward the doctors and nurses directing the treatment. Patients may also suffer feelings of guilt over earlier behavior, such as smoking or drinking, that might have caused the cancer.

Usually guilt feelings are unjustified. The many factors involved in causing cancer are so complex and so poorly understood that it is wrong to isolate a specific event or behavior in the past as its cause. Even with cancers of the skin and lung, for which sun exposure and smoking respectively have been clearly identified as powerful risk factors, remorse has little value at this point. A far better use of your energy would be to advise family and friends against taking comparable risks.

Finding Hope

An essential factor in coping with cancer is to remain hopeful and reasonably optimistic. For example, if you are dealing with a small skin cancer, your level of hope should be extremely high. If you have an advanced cancer of the esophagus or pancreas, then hope for reasonable comfort for as long as possible. It is essential that you and your doctor have a frank discussion as soon as possible, so that together you can establish realistic goals. If you need a clearer picture of your situation, ask someone you trust—your spouse, a close relative, or a friend—to talk to your physician.

Developing a Plan

Your ability to cope is greatly improved if you and your doctor map out a strategy or plan of attack. Be sure that you fully understand the steps involved in the diagnostic tests, treatment planning, the treatments themselves, and the follow-up visits. Ask your doctor or nurse for some informative brochures or booklets about your particular type of cancer. Patients and loved ones should read the brochures and feel free to ask questions about any uncertainties that remain.

It is important for you and your doctor to determine how far your cancer has spread. Many of your tests, such as bone scans, are done for this purpose. Determining how much your cancer has spread is called *staging* your cancer. Little or no spread is called an early stage—one or two. Extensive spread to other organs is called stage three or four. Your chances of a cure are much better at an early stage.

If no evidence is found that the cancer has spread, then it is reasonable to invest whatever is necessary in time, money, and discomfort for a good chance of cure. If cancer is found to have traveled to several spots in the body, however, then the probability of cure may be remote. Your doctor may recommend more gentle treatments that will *palliate* or relieve the symptoms of cancer without causing many side effects.

Developing a plan.

What Is a "Cure"?

What do doctors mean by the word *cure* with reference to cancer? Although the general meaning is the killing or removal of all cancer cells, no one can be absolutely certain of that. Therefore, one way of determining if a patient is cured is to note whether or not the patient has survived five years after the initial cancer treatment. Recurrences are rare after five years.

A *cure rate* could be calculated simply by finding the percentage of a particular group of cancer patients who are alive five years after their initial diagnosis. Some doctors count only the patients who are alive and free of all evidence of cancer. Others count all patients still living, whether or not traces of cancer remain. In some centers the survival rate is adjusted upward to take into account deaths that would be expected in the same age group from accidents and other causes. Thus, it may be difficult to compare the cure rates among various cancer centers.

Grief

Depending on the information given to you in your consultations with your doctor, you will undergo some level of grief. Grief is often misunderstood in our society; we mistakenly assume it is a sign of weakness. Yet it is a natural emotion, and one that the cancer patient and loved ones should express fully. Grief, after all, is a statement of our humanity, not of our weakness. It helps if family, friends, and physician all appreciate this.

During the period of grief it is essential for the patient to preserve three things: emotional balance, self-image, and relationships with family and friends. Emotional balance will give you a sense of perspective and will keep you from despair. It is normal and healthy to feel and express your grief, but don't let it become your master. Many patients are pleasantly surprised to find an emotional balance coming from an inner strength they didn't know they had. They may never have needed to draw from it until now. Many people find prayer of enormous value in preserving emotional balance.

How Do You See Yourself?

A positive self-image is extremely important in maintaining relationships with family and friends. Have you ever tried to project your mind's eye outside your body and look back at yourself from an outer perspective? How do you feel about the image you see of yourself? Do you sense a wall of reserve or defense, or do you see a person who is honest and open? You will feel much better about yourself if you see the latter image, and your family and friends will find it much easier to relate with you. They are probably feeling uneasy about discussing the disease with you, but they need information about your condition, too. They may be looking for a signal of openness from you. In the long run, everyone will feel better if there are no areas of forbidden conversation.

The Role of Family and Friends

As you begin to feel vulnerable, relationships and a sense of belonging with your family and closest friends become more important to you. If you think that you will ultimately die from this cancer, you should tell them. They may protest, thinking that you just want reassurance. If, however, you really want to talk about how you would like things to proceed within your family after you have gone, be firm about it. Eventually your loved ones will understand that this discussion is really important to you. Alternatively, if you would rather reminisce about the past, make that known, too. It can be very therapeutic to relive parts of your life with your family or friends.

Sometimes it's refreshing to make new friends, especially those who are experiencing similar challenges. Support groups called "One Day at a Time" or "I Can Cope" are usually active near major cancer centers. Your nurse or doctor can often put you in touch with the group's coordinator in your area. Sharing information, ideas, and feelings with other cancer patients is often very rewarding. It may lead to a more positive outlook and may actually extend your life.

In the mid-1970s a psychiatrist at Stanford University began an evaluation of the short-term effects of group therapy on patients with advanced breast cancer. Those patients randomly assigned to group therapy became less anxious, less fearful, and less depressed than those not assigned to group therapy. Through self-hypnosis the group members also learned to reduce their pain. Ten years later, skeptical of popular psychiatry programs claiming to conquer cancer through positive thinking, the psychiatrist decided to follow up on his earlier study. To his amazement, he found that patients who received a year of group therapy lived, on the average, twice as long as those who did not! This is not to suggest that psychotherapy is a substitute for proven medical treatments, but this small study of 86 patients certainly encourages people to think positively.

CHAPTER 3

Whom Should You See for Advice about Diagnosis and Treatment?

All Doctors Are Not Equal

Unfortunately, not all doctors offer equally good treatment. They differ in their personality, training, and experience. The most important qualities for a patient to look for in a doctor are competence, experience, warmth, and understanding. Your family physician or internist may have these qualities when it comes to dealing with illnesses common to the family and may help you through this early, trying period. He or she may not, however, have the specialized knowledge and experience to offer you the best cancer care. Perhaps your family physician has already suggested the name of a surgeon or other cancer specialist for you to see.

Sources of Information

Now is the time for you to get more information. Do you know anyone who is a cancer specialist? If you do, call that person right away and ask for the names of good cancer specialists near you. The

Cancer Information Service is another very important—and free—source of information. Just pick up your phone and dial 1-800-4-CANCER (1-800-422-6237) Monday through Friday. A recorded message will instruct you to touch "1" on your touch-tone phone to be connected with someone who can answer your questions. This person will be a highly trained nurse or social worker who has, right at his or her fingertips, a great deal of information about your problem. This service, supported in part by your tax dollars, cannot recommend specific cancer specialists to you. The staff can send you publications and tell you a great deal about your particular type of cancer. They can also tell you which types of specialists are most often involved in the diagnosis and treatment of cancer, and the names of the specialists in your area.

The Cancer Information Service (CIS) can also tell you about the latest clinical research studies being performed at major medical centers across the country for your particular type of cancer. They can tell you whom to contact to learn more about these studies. Ask the Cancer Information Service to look up this information for you from the Physician Data Query (PDQ).

Getting Another Opinion

What if you have already seen your doctor, and he or she suspects cancer and has referred you to a surgeon? See the surgeon without delay. Remember that you are asking for advice only, and that you are entitled to another opinion before you agree to any procedure (most insurance companies pay for two or three opinions).

If the surgeon recommends a biopsy, have it done. This will establish the diagnosis. If the biopsy is positive for cancer and your surgeon recommends major surgery, you are under no obligation to follow that recommendation. You should obtain a second opinion from a cancer specialist, not from a colleague in the same office. The second opinion could be from a radiation oncologist or a medical oncologist. (See Chapter Five or the Glossary for definitions of these terms.)

It is often best to work from the beginning with a team of cancer specialists. For example, your radiation oncologist should have an opportunity to examine the cancer before the surgeon removes it. The radiation oncologist can then better plan your course of radiation therapy.

Sometimes radiation therapy and/or chemotherapy given before surgery can make the operation easier to perform, especially if the radiation therapy or chemotherapy significantly shrinks the tumor. Finally, both radiation and medical oncologists have had extensive training in how certain types of cancer spread. They are often able to assist your surgeon in requesting the various tests to determine the stage of the cancer and choosing the optimal treatment plan.

Testing and Waiting

As soon as your doctor is certain you have cancer, you might be eager to start the treatment as quickly as possible, perhaps due to fear of the cancer spreading or because you want to get the treatment over. In order to tailor the treatment for you and your cancer, however, it is essential that your doctor order additional tests, especially if the tumor is deep within your body. Such tests may include various x-ray studies or "scans." It may take a week or two for your doctor to arrange for all of the tests and to study the results. Therefore, try to be patient; your cancer is unlikely to spread within a week or two, and your treatment will be only as good as the knowledge that your doctor has about your particular cancer.

Talking to Your Doctor

Finally, try to maintain an open relationship with your doctor. Although some patients would rather not know that they have cancer or what the prognosis is, most patients do want to know as much as they can about their case. In return, doctors respond more to patients who clearly express this need.

If you are seeing the doctor on behalf of a child over 10 years old or an elderly parent, think long and hard before you ask the doctor not to tell the diagnosis or prognosis to the patient. First, you are asking that the doctor join you in a conspiracy of silence that would inevitably interfere with the doctor-patient relationship. Second, it is very rare that a patient—young or old—doesn't realize that something serious is happening when he or she has to stay alone in a room for several minutes under an enormous x-ray machine. Let your doctor help you to make the decision whether or not to tell the patient.

CHAPTER 4

Where Should You Go for Treatment?

People with cancer often travel to famous medical or cancer centers for treatment. Large centers have many cancer specialists on staff, the latest technological advances, and experimental therapy. If you live near such a center, you are indeed fortunate. If you live in a smaller community some distance away, however, you need to know whether the more complex care provided at a larger medical center or university hospital will make a difference in treating your cancer.

Most patients with curable cancers and essentially all children are best treated by the teams of cancer specialists found in a major medical center. Be aware, however, that university hospitals and major cancer centers do not offer miracles—there is just no effective therapy yet for some advanced cancers. It would be a pity to leave the loving support of your family and friends and spend money unnecessarily when comparable supportive care can be found close to your home.

Perhaps the best way to decide where to go for treatment is to ask your cancer specialists whether or not you can receive good care in your own community. If traveling presents a problem, ask the specialists if they can recommend a treatment center nearer home.

You can get additional help from the Cancer Information Service at 1-800-4-CANCER. If you ultimately decide that good care is available in your own community, then do stay closer to your home. The advantages of closeness to family and friends, and savings of time and money, are obvious.

Types of Cancer Treatment Centers

If good treatment is not available in your community, what should you do? The following brief descriptions of university hospitals, comprehensive cancer centers, and large medical clinics may help you decide which one is best for you.

University Hospitals*

Like major medical and cancer treatment centers, most university hospitals have an abundance of well-trained specialists in cancer treatment. An important difference from the large clinics is that the doctors in university hospitals train medical students and *resident* doctors who are in training to become cancer specialists. In a university hospital, these residents will also be directly involved in your care. Furthermore, you may not always see the same doctors at each visit. These differences bother some people, but others find them a minor issue because of the high quality of care available.

---

*a complete listing follows the Glossary

Comprehensive Cancer Centers*

Well-known examples of major cancer treatment centers are the Memorial Sloan-Kettering Institute in Buffalo, New York; M.D. Anderson Hospital and Tumor Insititute in Houston, Texas; and the University of Wisconsin Comprehensive Cancer Center in Madison, Wisconsin.

Cancer treatment centers offer the same advantages and possible disadvantages as large university hospitals. These major centers have teams of specialists who treat each type of cancer, so they are very experienced. The National Cancer Institute in Bethesda, Maryland, another comprehensive cancer center, strongly emphasizes experimental therapy. There are no fees for treatment if you are eligible for one of their studies.

Major Medical Centers

Some of the best known medical clinics in the United States are the Cleveland Clinic in Cleveland, Ohio, the Mayo Clinic in Rochester, Minnesota, and the Lahey Clinic in Burlington, Massachusetts. Specialists in these centers spend most of their time on patient care and clinical research, without the distraction of teaching or laboratory research.

---

* a complete listing follows the Glossary

Community Hospitals

Some of the larger hospitals in medium-sized towns also provide radiation treatment facilities. Many of these centers offer good quality radiation therapy. However, it would not be unreasonable for you to ask for certain reassurances, such as how modern are the simulator and linear accelerators, or is the medical physicist available full-time?

Free-Standing Radiation Centers

There has been a recent trend toward the construction of radiation therapy centers away from hospitals, often in suburban districts around major cities. Many of them are well staffed and have modern equipment. Of course, you are always entitled to ask questions about the quality of care that would be best for you.

More than half of all cancer patients are cured. To give yourself the greatest probability of being one of those cured, you must make certain you are getting the best possible treatment.

CHAPTER 5

How Is Cancer Usually Treated?

There are three main ways of treating cancer:

- **Surgery**
- **Radiation therapy** (killing the cancer cells with radiation, such as high-energy x-rays)
- **Chemotherapy** (giving chemicals or drugs that keep the cancer cells from dividing to form more cancer cells)

In recent years two new approaches are being studied—*immunotherapy* (helping the body's immune system kill "foreign" cells) and *hyperthermia* (applying heat to the cancer cells and artificially raising their temperature). We will discuss these in Chapter 9.

Each method of treatment has advantages and disadvantages. In a modern cancer center they are often used in combination. The method(s) used depends on the type of cancer, where it is located in the body, and how far it has spread. Each method alone can cure particular types of cancer, especially in its early stages. Also, any one of these methods may be used to reduce suffering and prolong useful life when the patient's cancer is incurable by any method. This is called *palliative therapy*.

Surgery

Surgery for cancer was used long before radiation therapy and chemotherapy were even discovered. The goal of using surgery to treat cancer is to be sure all cancer cells are removed. That's why many cancers are removed in what's called a *radical* operation. This involves removal of the entire primary tumor, any tissues nearby that might contain microscopic extensions of tumor cells, and as many lymph nodes that drain the tumor area as possible.

In spite of dramatic improvements in radiation therapy and chemotherapy over the past two decades, surgery still cures more cancer patients than radiation and chemotherapy combined. This is because many cancers are detected at an early stage, and surgery alone often produces a cure. Recent improvements in surgical techniques, anesthesia, and postoperative care have also made surgery safer and easier than it used to be. In addition, some surgeons specialize in cancer surgery or even one branch of cancer surgery—such as surgery of cancer of the brain.

If you are considering surgery for your cancer, ask your doctor the following questions:

- Is my cancer still at a stage that is likely to be cured by surgery?
- Is surgery the best initial treatment?
- Are there alternative treatments that are as effective as surgery?
- What parts of my body (e.g., muscles, nerves, etc.) might need to be removed along with the cancer?

- Are other forms of treatment likely to be recommended after surgery?
- What are the chances of the cancer regrowing after surgery (if no treatment is used in conjunction with surgery)?
- What are the chances of my cancer spreading?
- What side effects should I expect from this surgery?
- How often does the surgeon perform this type of surgery, and what is his or her success rate?

If these questions are answered to your satisfaction, then you will have more confidence in your surgeon. Similarly, if your surgeon performs many operations every year of the type planned for you, he or she is more likely to perform expertly. Look at the diplomas on the wall when you enter the surgeon's office, too. If he or she is a member of the American Society of Surgical Oncologists and the American College of Surgeons, then you can have more confidence in the surgeon's ability.

You may be asked to take part in deciding which type of surgery should be performed. If you have a small breast cancer, for example, your surgeon should inform you that removal of the breast lump followed by radiation therapy is just as effective a treatment as removal of the entire breast.

Radiation Therapy

Unlike surgery, x-rays destroy cancer cells without actually removing them. The two forms of treatment often work together. Surgery removes the main bulk of cancer cells, while x-rays kill microscopic extensions of the cancer that the surgeon either couldn't see or couldn't cut out because of their location. In this situation, the patient enjoys three advantages. First, this combined approach often reduces the chance of a cancer recurrence. Second, it may allow the surgeon to spare more normal tissues. Third, it may allow the *radiation oncologist* to use slightly lower doses of x-rays, which are usually well tolerated by most normal tissues. Some examples of cancers that are often treated by this combined approach are cancers that start in the throat, lung, breast, rectum, or uterus.

If detected early, cancers of certain organs that are very precious to us, such as the vocal cords, are often treated with radiation alone. The patient receives x-ray treatments to his voice box (larynx) for a few minutes every day (Monday through Friday) for approximately six weeks. At the conclusion of the treatments the tumor has usually disappeared completely. The patient may have some skin inflammation over a two-inch area in the Adam's apple region and some hoarseness. He would be told not to use his voice excessively for a few weeks; the sound of his voice usually returns to normal after a month or two.

Other cancers, such as lymphomas, are very sensitive to radiation. X-rays are often the only type of therapy needed to treat Hodgkin's Disease, for example, even though large regions of the body need to be treated.

Even certain cancers of the eye can be treated with radiation alone. In this case, very low-energy x-rays from a radioactive substance are used. The radioactive sources are embedded in a plastic disc that has the same curvature as the patient's eyeball. The disc is sewn onto the outer surface of the eye while the patient is anesthetized. When the patient wakes up there is minimal discomfort. The disc stays in place over the eye tumor for almost a week, allowing the radioactive substance to treat the cancer continuously with x-rays. After therapy is complete, the disc is removed under local anesthesia, and the tumor usually regresses very slowly during the next year.

Radiation therapy for breast cancer is somewhat controversial. The general public is used to the idea that mastectomy is the preferred treatment because of the choices made by two presidents' wives, Betty Ford and Nancy Reagan. Extensive research in Europe, Canada, and the United States has shown, however, that surgical removal of a small cancerous lump followed by breast irradiation is just as effective as surgical removal of the entire breast.

X-ray therapy is also used to treat cancer that has spread—especially if the patient has pain or bleeding from the metastases. With rare exceptions, patients with metastatic cancer are not presently curable. Treating them with high doses of radiation would take too much of the patient's time and risk unnecessary side effects. Since a cure is not possible in these circumstances, radiation is used only to relieve symptoms. Therefore, treatments take only two to four weeks and are designed to arrest or shrink the cancer growth for several months or even a few years.

An exception to this would be a patient with a single metastatic nodule of cancer, where the original (primary) cancer was successfully treated months or years previously. There is a significant (30%) chance that such a patient can be kept free of cancer for several years and perhaps even cured if the single nodule of metastatic cancer can be eradicated permanently. Such patients may be treated for five or six weeks in the hope of curing the cancer.

There are two good features about x-ray therapy for metastatic disease. First, it is usually effective (about 70% of the time) in relieving most symptoms, especially pain. Secondly, it is usually well tolerated with few side effects, especially if the area to be treated is relatively small. (See Chapter 7 for details.)

Unfortunately, x-rays only affect the cancer cells that they pass through, so any cancer cells in unirradiated parts of the body continue to grow. If the size of the x-ray beam were enlarged to cover the entire body, conventional doses of x-rays would do too much damage to normal cells. This is why chemotherapy is the preferred method of treatment for metastatic cancer that is located in many parts of the body.

Chemotherapy

Chemotherapy is the treatment of cancer with chemicals—medicines—designed to kill cancer cells or halt their spread without causing serious damage to healthy cells. There are many types of chemotherapeutic drugs. They fight cancer either by preventing cells from dividing or by preventing cells from getting the nutrients they need to survive. Chemotherapy may be taken by mouth, *intravenously,* or injected into the body, such as the lung or abdominal cavities.

Doctors have dreamed for decades of developing drugs that would kill only cancer cells and leave normal cells unharmed. Unfortunately many of the drugs developed so far do affect normal cells temporarily. They may cause side effects such as hair loss, sores in the mouth, nausea, vomiting, diarrhea, and depressed blood-cell counts.

Chemotherapy drugs are not effective against all types of cancer, and some cancers eventually build up a resistance to these drugs and grow again. When this happens, medical oncologists usually change the prescription to a different drug, against which the cancer has not had a chance to develop any resistance. Often several drugs are given at once or in a precise sequence over a treatment cycle of three or four weeks in order to kill as many cancer cells as possible. These treatments continue for as long as the patient can safely tolerate them or for as long as the cancer seems to be responding to the therapy. Usually this is less than a year.

Sometimes chemotherapy is given to patients who currently do not have any detectable cancer but who have had it before and have a high risk of getting it back. This treatment, called *adjuvant chemotherapy*, has benefited some patients who have had breast cancer, especially if the cancer spread to the lymph nodes under the arm.

Chemotherapy and radiation therapy are usually not given at the same time. For certain cancers of the throat, esophagus, or anal canal, however, doctors may recommend that some of the chemotherapy be given on the same days as certain radiation treatments. This may increase the likelihood and severity of side effects, but the advantage may outweigh the discomfort: cancers sometimes shrink so completely after this combined therapy that surgery is not needed.

Some breast cancers require female hormones in order to grow, and some prostate cancers require male sex hormones in order to grow. In these situations the doctor will deprive the tumor of the hormones it needs. This may be done by removing the ovaries or testicles. Another approach is to use a drug that binds to hormone receptors on the tumor cells, blocking the tumor's access to hormones present in the body. This may cause the tumor to stop growing or even shrink. The encouraging aspect of this *hormonotherapy* is that it is generally well tolerated by most patients, and the tumor will stop growing for many months or even years.

So far I have introduced you to the three most important forms of cancer treatment: surgery, radiation therapy, and chemotherapy. You will find further details on radiation therapy in Chapters 6, 7, and 8. Newer types of cancer treatment such as hyperthermia and immunotherapy are described in Chapter 9.

CHAPTER 6

Radiation Therapy for Treating Cancer

The Role of the Radiation Oncologist

There is a misconception among the general public that radiation therapy simply involves aiming x-rays at a patient's cancer. The truth is that modern radiation is performed by a cancer specialist with three or more years of specialized training in addition to his or her training as a medical doctor. This training includes: (1) how each type of cancer spreads, (2) the effects of x-rays on normal and cancer tissues, and (3) the physics of radiation therapy. With that knowledge at hand, the radiation oncologist is an important partner with the referring physician. The two doctors can decide together which tests to run in order to determine if or how far the cancer has spread. These tests might include a *bone scan*, *CAT scan*, or *MRI scan*—all special tests that show the location of organs and tissues deep within the body. The scans help the doctors check for the spread of the cancer to the bones, soft tissues, or other regions of your body.

Once the cancer is identified and staged, you and your radiation oncologist will want to discuss the results of these tests and set realistic goals. Ask your doctor for a summary of what is known about your cancer, and any additional information that may be needed before treatments begin.

It's a good idea to bring a family member or close friend along to this appointment to help you with any decisions you'll need to make. There is often much to remember, and your companion may be of great assistance in helping you to sift through the less important material and focus on the most important issues.

Bring a family member or friend to help you sort through the information your doctor gives you.

The radiation oncologist will usually outline a tentative treatment plan. This plan includes the area or volume to be treated, the possible construction of special supports to help you to lie still during each treatment, the number of treatments and their frequency. He or she will explain to you possible side effects, and the overall goals of therapy such as pain relief or possible cure. You should be given rough estimates of the likelihood of each side effect and the probability of curing your cancer.

At this point you should discuss your doctor's plan. Does it sound reasonable to you to undertake the outlined risks for the anticipated goals? Ask any questions that might help you to decide on whether or not to go ahead with the treatment plan.

Some questions you might ask include:
- What is the likelihood that radiation will make this cancer go away permanently?
- What are the chances that these radiation treatments will shrink the cancer and make the pain (or other symptoms) stop?
- What side effects should I expect from x-ray treatments? What are some of the less likely side effects?
- How will I be positioned during the treatments? Where will the x-rays enter my body?
- Could these x-ray treatments cause another cancer later?
- Will the treatments have an effect on any other part of my body?
- What alternative types of therapy are available?

By the time you have read this book you may feel that you know some of the answers already, but ask your radiation oncologist anyway. It may reassure you to hear the answers from him or her, and the discussion will help the two of you to establish a relationship.

Sometimes your decision will be easy, especially if the risks are small and the potential benefits great. But the decision isn't always a simple one, and you may want to go home and think it over for a few days.

When you and your radiation oncologist agree upon a plan of therapy, you will probably be asked to sign a consent form. Read it, and sign it if you understand it and agree with the statements. It will not bind you to a specific number of treatments. Your signature on the consent form indicates that you understand the potential risks and benefits of treatment and that you agree with the plan as outlined.

Planning Your Treatment

The goals of radiation therapy are to concentrate the radiation into the tumor and minimize the radiation to the nearby normal tissues. Accomplishing this goal requires the use of special equipment such as treatment simulators (specially designed diagnostic x-ray units) and computers. Planning your radiation therapy requires the services of specially trained personnel —dosimetrists and medical physicists.

After you have signed the consent form, your doctor will arrange for you to meet with the scheduling secretary. He or she will tell you when the next appointments are available for a *CAT scan* and *simulation*. This is usually within a few days. These procedures and all of your treatments will generally be carried out in the radiation therapy department, usually in the basement of a hospital. You can be treated as an outpatient (coming to the hospital for treatment, then leaving when the procedure is completed) unless your medical condition requires you to be hospitalized.

A CAT scan and simulation may each require one to two hours. The success of your treatments depends on how precise the CAT scan and simulation are. You can help by lying as motionless as possible during your scan and simulation. Some patients find this the

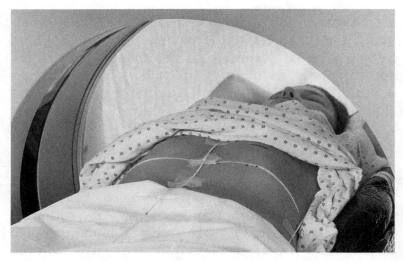

In the CAT scanner. Before the actual scan, your technologist may tape thin tubing onto your body. This tubing will show up in the images and relate surface marks to the deeper regions within your body that need to be treated.

most difficult part of therapy because lying still can be quite uncomfortable if you have pain. If this is a problem for you, mention it to your radiation oncologist so that he or she can prescribe enough pain medicine for you to be reasonably comfortable during these very important planning procedures.

At this stage you will first meet one of your radiation therapy technologists (also called "radiation therapists"). The technologist is trained to accurately deliver the dose of radiation prescribed by your doctor. The technologist is also responsible for helping you to get into a comfortable position that will be best for letting x-rays pass through the part of your body being treated. He or she will provide special cushions or foam supports so that you can maintain the same

A Cancer Patient's Guide to Radiation Therapy - 33

position for every treatment. You may also meet your medical physicist. A medical physicist advises on technical aspects of the treatment plan.

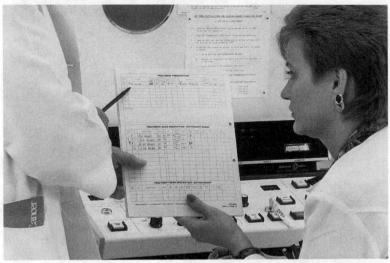

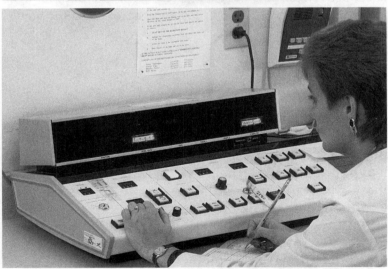

The technologist reviews the oncologist's prescription, and then inputs the data into the treatment console.

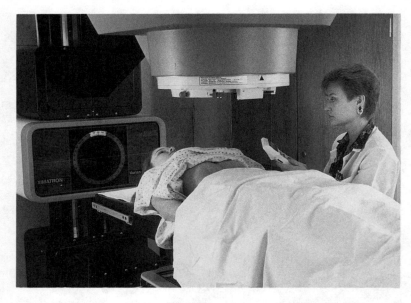

Under the simulator. The simulator is a machine somewhat like a linear accelerator. It is designed to mimic, with a beam of light, the radiation beam delivered by the linear accelerator. The simulator does not deliver radiation.

Being Simulated

Some patients are confused by the notion of being *simulated*. Just as a flight simulator allows a student pilot to experience most aspects of flight without leaving the ground, the x-ray therapy simulator helps the technologist and radiation oncologist design your treatment before you are given any therapy. The simulator allows the technologist and radiation oncologist to see clearly what part of you will be treated by taking conventional x-ray pictures before the actual treatments begin. They can see the part of your body with cancer, plus a rectangular outline of the planned treatment area (called the *treatment field*) on a special x-ray instrument called a *fluoroscope*. While standing in a shielded area they can adjust the position of the treatment table on which you will be lying, and select the angle of the x-ray machine and the size of the x-ray beam needed to include all of your cancer. An x-ray picture is taken as a permanent record of the treatment field. After the x-ray film has been developed, your doctor will evaluate it to make certain that the treatment field is exactly the way he or she wants it to be.

During treatment planning and simulation the oncologist or technologist will place marks on your skin to aid in the positioning of your body and precise aiming of the x-ray beam for every treatment. Technologists often use a purple dye, called *carbolfuchsin*, (pronounced "car-bo-fyu-shun") for this purpose because it is more resistant to washing off than the ink used in most marker pens. Continue to shower, using a mild soap such as Dove or Ivory. Allow the water to run over irradiated skin and these marks, but do not scrub them with a washcloth. Gently pat the area dry with a towel you don't care about, because some of the dye will stain your towel. The dye will wash off after a week or so. Occasionally your technologist will add more dye to your marks to prevent them from disappearing during the treatment period. If some of the marks will be visible, especially on your neck or face, ask your doctor and technologist about keeping the marks to a minimum. Sometimes a small black dot can be tattooed into the skin as a permanent marker. (The latter method is preferred by some hospitals, especially to verify radiation borders over an important normal structure such as your spinal cord.)

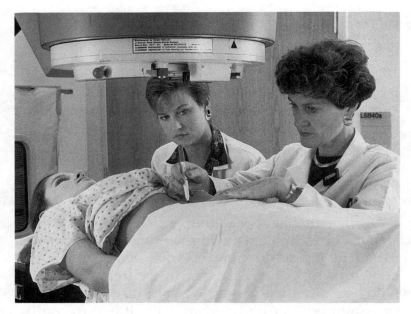

Placing the marks.

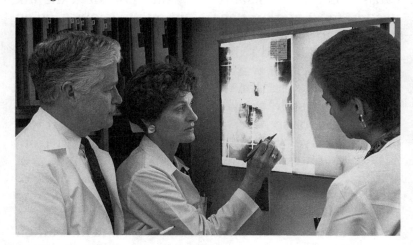

Evaluating the treatment field.

Your doctor may draw lines with a wax pencil on the x-ray so that special lead blocks can be designed for protecting sensitive parts of your body from the radiation. These protected areas might include your spinal cord, lungs, heart, or kidneys.

Meanwhile, the technologists use visible light to project the center and borders of the area that will be treated onto your skin. They mark these important lines with the dye markers previously described. If the x-rays will be aimed at the tumor from several different angles, it will be necessary to repeat the simulation from each of those different angles. The entire procedure may take over an hour.

Your First Treatment

Usually, your treatments won't begin until a day or two after simulation because it takes this long for your lead blocks to be made. On the day of your first treatment, wear comfortable clothes that are easy to slip on and off. If your ears or neck will be irradiated, don't wear earrings or necklaces. Don't worry if you forget to leave your jewelry at home. Most radiation therapy centers provide cubicles and lockers for removing and storing your belongings. Hospital gowns are, of course, provided.

Be sure to arrive on time; you won't want to keep others waiting because you were late. In a larger radiation center there may be 100 or more patients treated every day, and you'll want the staff to know where to find you. Check in at the front desk and at the treatment unit.

Check in at the front desk before every appointment.

Occasionally, you may be kept waiting because the high-energy x-ray machine (linear accelerator) to which you have been assigned isn't working that day. Sometimes the department's engineer can fix it promptly, but it may be necessary to switch you to another linear accelerator and to fit your treatment in among the patients already scheduled for that machine. This can be frustrating, but fortunately it does not happen very often. Linear accelerators are inspected regularly and are kept in peak operating condition, but it's wise to bring a good book along with you, just in case.

In the Treatment Room

Radiation treatments are usually given once a day, Monday through Friday. Only for rapidly growing cancers will your radiation oncologist recommend two treatments a day or treatments on the weekend.

The x-ray treatment itself only takes a few minutes. You will be asked to lie down on an adjustable treatment couch. A highly trained radiation therapy technologist will help you to get into a comfortable position, which you will be asked to maintain during the treatment. If the staff sees that you need support for your head, arms, or knees, they will provide it for you. The part of your body through which the x-ray beam will pass will have to be exposed so that the technologist can be sure that you and the linear accelerator are lined up correctly. If you feel that you are unnecessarily exposed, ask your technologist to cover part of you with a small towel.

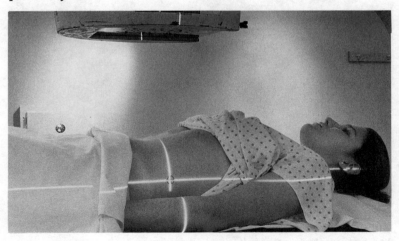

On the linear accelerator. The lights are aligned with the marks that were made on your skin during simulation.

The couch will rise slowly and approach the linear accelerator. Don't be afraid; you won't fall off the couch, and the machine won't touch you. The technologist will darken the room a little, and use precisely aimed red lights, called *lasers*, to make sure that you and the treatment couch are in exactly the right position. If the x-rays are prescribed from an angle other than directly above you, you might see the big arm, called the *gantry* of the linear accelerator, move around you. You might also notice on the part of the linear accelerator closest to you a clear plastic Lucite tray with lead blocks attached to it on the end of the linear accelerator. The lead blocks have been specially designed to shield the parts of your body that do not need to be treated.

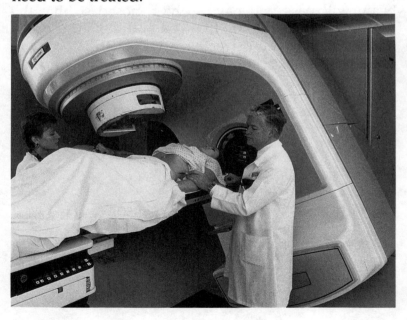

The accelerator's arm, called the gantry, moves around you as you lie on the treatment couch.

During your first treatment a *dosimetrist* may come in to calculate the amount of time needed to deliver the dose of radiation prescribed and possibly suggest minor modifications to the treatment plan. A medical physicist may also check the treatment at this time, just to be certain that the treatment plan is being carried out precisely. Your radiation oncologist may visit you, too. You might feel that you are the center of a great deal of attention, and wish that the treatment would start more quickly. Try to be patient and remember that if the first treatment is given exactly as it should be, subsequent treatments are likely to be given correctly, too.

Once everything meets with the specialists' satisfaction, they will all leave the room. They will turn on a radio or tape recorder, if you like, to provide some pleasant background music. The linear accelerator may make a clicking noise occasionally, but try to ignore it and breathe quietly, remaining as still as you can.

On the first day of your treatment, and perhaps weekly thereafter, the technologists might take *check films* to make certain that they have reproduced the treatment setup exactly. This could add an extra 15 to 30 minutes to your time in the treatment room. The actual time that the linear accelerator is treating you each day, however, is only two or three minutes.

After your treatment has been given, the technologists will come back into the room and lower the treatment couch. They might ask you to wait for a moment while they add some dye to the markings on your skin. Then they will help you up and into your clothes or to a changing room. In all probability you will not feel any different than you did before the treatment, but side effects as well as good effects from treatment may become evident after a few weeks. (Side effects will be discussed in Chapter 7.)

Radiation oncologists often prescribe slightly higher doses of x-rays today than they did 20 years ago, with better cancer-killing success. Patients generally encounter fewer side effects than 20 years ago, largely as a result of the availability of better treatment machines and greater precision in treatment planning.

Patients often want to complete their therapy as quickly as possible, and occasionally they ask why it takes so many days or weeks to complete radiation therapy treatment. The duration depends upon the goals of the therapy. Treatments can vary from a single treatment to inhibit normal tissue growth near a scar, for example, to three months or more in an attempt to cure a patient with Hodgkin's disease. These are extreme cases. Radiation therapy to reduce pain in a patient with cancer that has spread to bone is often delivered in two to three weeks. Radiation therapy given after cancer surgery to kill the few cancer cells that might have been left behind often requires five or six weeks. During treatment most patients continue to live at home. Many continue with their work, and drive to the hospital each day for their treatment.

Electron Beam Therapy

During the last week or two of therapy your radiation oncologist may prescribe a change in your treatments from x-rays to electrons. Sometimes this is part of the overall plan to shrink the volume of tissue being irradiated and destroy any lingering cancer cells by focusing on the region of greatest importance. In contrast to x-rays, electrons can be made to travel a short distance in your body (one to three inches, for example), and then stop rather abruptly. This can be very useful for treating certain superficial parts of the body that lie over sensitive deeper structures (such as

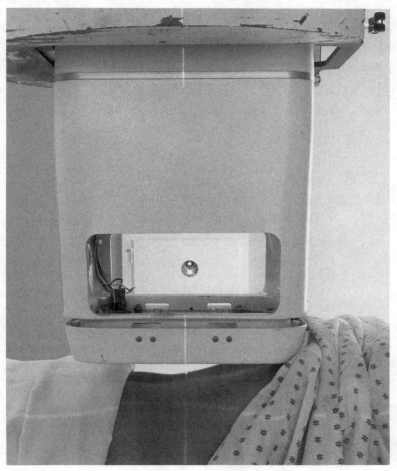

Electron beam therapy.

lung or spinal cord) that must be protected.

The treatment planning for electrons is relatively simple, and may not require a simulation. If your radiation oncologist simply outlines the electron field with a felt-tipped pen, a dosimetrist can make a special lead shield to match it within a few minutes. This lead shield fits onto the linear accelerator with a special device called a cone, shown in the figure on page 43. It will come very close to you, but your technologists will be careful not to let it press against you too firmly.

CHAPTER 7

What Are the Side Effects of Radiation Therapy?

It is quite natural for you to be concerned about the possible side effects of x-ray treatments. The side effects you may experience, if any, depend greatly upon your individual case. What portion of your body and how much of it will be irradiated? What is the dose of x-rays used? How many treatments were prescribed over how long a time period?

Side Effects During Treatment

In the relatively rare situation where a large dose of radiation is given to the entire body to suppress the immune system before a bone-marrow transplant, the patient may experience nausea within minutes, and diarrhea within hours of the treatment. In contrast, the much lower doses of whole-body irradiation used to control certain lymphomas don't produce nausea or diarrhea at all.

Usually, however, x-rays are prescribed for only a small part of your body, where the cancer is located. If the dye marks on your skin outline an area that you think is larger than your cancer, remember that many cancers can spread along fine lymphatic channels to nearby lymph nodes. Your radiation oncologist may think that it is appropriate to treat those lymphatic channels or nodes as an extra precaution. In any case, early side effects will usually take two or three weeks to develop, and they will relate to the part of your body being treated.

An exception to this is fatigue (tiredness), which may develop within the first week of treatment and seems to be related more to the amount than the location of irradiated tissue. It will persist until the treatments are completed, so plan to take more rests during your course of treatments.

Another side effect to expect is redness of the skin in the treated area. You may also experience a slight swelling and/or tenderness in the soft tissues just under the skin. This might occur after three to four weeks of treatment. Deeper *mucous membranes* such as those in the lining of the mouth, throat, esophagus, stomach, or bowels might become sore a week sooner if they are also irradiated. You might feel a lump in the esophageal lining when you swallow, or have diarrhea if the bowels are irradiated. If your mouth, throat, or esophagus are being irradiated you can reduce the severity of these side effects by avoiding food containing acids. If your bowels are being irradiated, avoid high-residue cereals or gas-producing vegetables such as legumes or cabbage. Let your radiation oncologist know about any side effects you're experiencing so that he or she can make adjustments to your treatment plan as necessary.

Hair loss and nausea, two possible side effects from irradiation, are not as common as most people seem to think. Hair loss occurs at three to four weeks, but only if radiation is delivered to the brain, scalp, or the upper back of the neck. Unfortunately, nausea is not as easy to predict because it varies so much from one person to another. Doctors generally believe that irradiation of a considerable length of the spinal cord is more likely to cause nausea than irradiation of a smaller region. There is an enormous range of sensitivity in this regard. Some individuals are very tolerant of x-rays while others develop nausea after little or no spinal cord irradiation.

Post-Treatment (Late) Side Effects

The early side effects will usually persist only one or two weeks after the course of treatment is over and then gradually subside. As the discomfort fades, some or all of the functioning cells in the irradiated area will gradually be replaced by diffuse scar tissue. This leads to the relatively permanent radiation effects in the treated area. For example, sweat glands in the treated area, especially those in the armpit, will dry up after one to two months. If you lost hair from a treated area, it may not be quite as thick when it regrows as it was before the therapy.

To minimize damage to sensitive and vital organs such as the lungs or liver, radiation oncologists keep the irradiated areas as small as possible. Damage to the irradiated parts of these organs can usually be detected by x-rays or special scans taken months later. Irradiation for certain cancers of the throat may include your salivary glands, which will probably lead to reduced flow of saliva and to a relatively dry mouth.

Some patients ask why the course of radiation treatments cannot be shortened for their convenience. "Instead of six weeks," they ask, "why can't I receive the radiation I need in three weeks?" The answer is that if the x-ray treatments are given in a shorter time, larger doses must be given each day and the chances of side effects go up.

In general, patients treated with the hope of a cure are more likely to experience both early and late effects of radiation. Higher doses are used and the normal tissues near the tumor are treated to the same high doses as the tumor. Most patients are willing to accept the greater side effects in order to improve the chances of having their cancer cured.

Radiation-Induced Cancer

There is one effect of radiation that is fortunately very rare: a new cancer many years later. Normal cells irradiated during treatment may be altered into cancer cells. The change from a normal cell to a cancer cell may take many years. Atomic-bomb survivors who received high levels of radiation had an increase in cancer many years later. This was especially true for children. Of the 100,000 bomb survivors there were about 340 excess cancer deaths in the following forty years.

With recent advances in surgery and chemotherapy for the treatment of childhood cancers, radiation therapy for children is used less often today than it was 10 or 20 years ago. When x-ray treatments are given to children, the dose and the area treated are kept to a minimum. The risks of x-ray treatments are carefully weighed against the potential benefits. Since the time interval until radiation-induced cancers appear may range from 5 to over 20 years, this rare but very serious complication is taken more seriously in children than in elderly cancer patients. It has to be balanced, however, against the knowledge that trimming the dose or the field size too much in children will reduce the chances of curing the cancer.

CHAPTER 8

Special Types of Radiation Therapy

Brachytherapy

There are two different ways of delivering radiation to kill cancer. Until now we have been considering the most commonly used way, called *teletherapy* (in Greek, *tele* means far off). During teletherapy the x-rays are made in a machine that is about three feet away from the tumor. The dose given to the tumor volume and the tissues around it is relatively uniform.

The other way to treat cancer with radiation is called *brachytherapy* (in Greek, *brachy* means short). With this technique radiation sources are placed very close to the tumor, sometimes in direct contact with the tumor, for a few days. These sources emit radiation continuously.

Shortly after radium was discovered early in this century, it was placed in the vaginal cavity of patients with cancers of the uterine cervix with considerable success. Later, radium needles were manufactured. These could be implanted into cancers in the mouth and many other sites.

Radiation administered in this manner is confined mostly to the cancer itself. Brachytherapy is a good way to increase the dose of radiation to the cancer without causing excessive damage to normal tissues around it. Until recently the advantages of this treatment method were not widely appreciated for cancers other than cancer of the uterus. This is because modern brachytherapy techniques and safer radioactive sources which protect the radiation oncologist, physicist, and other medical personnel from overexposure to radiation are constantly being developed. In addition, computerized treatment planning has changed the placement of radioactive sources and the calculation of radiation dose in brachytherapy from an art to a reproducible science. In short, there is now a resurgence of interest in the use of brachytherapy for treating many cancers, including even cancers in the brain.

Brachytherapy is most useful for small, superficial, or readily accessible tumors. It can also be used in a natural body cavity such as the bronchus, esophagus, and vagina with an increased likelihood of permanent cure.

Brachytherapy is not a good choice for every cancer patient. First, a general anesthetic may be needed for placement of the radioactive source in the cancer. Second, some cancers are so close to vital structures such as major blood vessels, nerves, or lung tissue that implanting the radioactive source is too great a risk.

Stereotactic Therapy

Medical physicists with special training in radiation therapy physics have greatly improved the quality and precision of cancer treatment. A new approach in which this is especially evident is stereotactic therapy of the brain. *Stereotactic therapy* is a very specialized type of teletherapy which requires the use of a computer because of the complexity of the treatment planning.

In stereotactic therapy the head is secured rigidly to the treatment couch so that no movement will occur. Then the arm of the linear accelerator moves continuously around the head directing a narrow beam of radiation very precisely at the tumor's center.

This approach is most appropriate for cancers that are relatively small (less than an inch in diameter) and deep within the brain. Such tumors are difficult to treat with brachytherapy because surgical placement of catheters into deep brain structures would be very dangerous. Treatment of such tumors with conventional teletherapy would cause too much damage to normal brain tissue.

Intraoperative Radiation Therapy

There is yet another special way in which teletherapy can be used to achieve one of the goals of brachytherapy; that is, to limit the radiation almost exclusively to the cancer. For intraoperative therapy, the cancer patient is taken to the operating room, where an anesthetic is given. Then an incision is made to expose the cancer. As much of the cancer is removed as is feasible. Organs that are very sensitive to irradiation such as liver, lungs, or small bowel, are gently moved aside with sterile, warm towels. Then a cone-shaped tube is attached to the linear accelerator, carefully aimed at the cancer, and the residual cancer is irradiated in a relatively large, single dose. The incision is then sewn up, and the patient is taken to the recovery room.

Cancers that might be appropriate for intraoperative irradiation are those that are deep within the abdomen or chest, especially if they are growing into vital structures that cannot be surgically removed. This type of therapy is usually combined with conventional external teletherapy either before or after the operation, so that the patient can receive the advantages of both approaches.

CHAPTER 9

Other Types of Therapy

Both hyperthermia and immunotherapy can by themselves kill cancer cells, but they usually work better when combined with x-rays and/or chemotherapy. At this time they are only prescribed for certain types of cancer, but in the future they may be used more often.

Hyperthermia

Hyperthermia means raising the tumor's temperature. The basic idea is to heat the cancer by 10 or 11 degrees Fahrenheit above normal body temperature while keeping the heating of normal cells to a minimum. This is usually done just before or just after radiation is given. If all of the tumor can be heated to this temperature for one hour, the x-rays will kill about twice as many cancer cells in a heated tumor than in an unheated tumor.

A superficial cancer can be heated by placing a microwave or ultrasound generator directly over it. This treatment, called 'local hyperthermia, might be used for a breast cancer that has recurred on the chest wall after mastectomy. Unfortunately, many cancers are not close enough to the surface to be heated by this means. Several approaches are used to deal with this

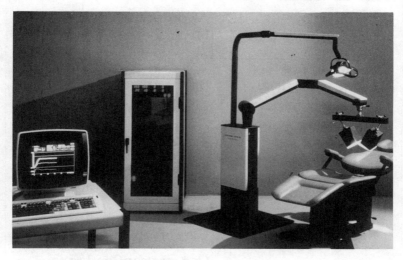
Hyperthermia treatment equiment.

problem. One, method, called *regional hyperthermia*, is to use many heating elements around the body to raise the temperature of the part of the body containing the cancer. Another is to heat the entire body by just 8.6° F, the maximum level that can be tolerated safely. A third approach, called *interstitial hyperthermia*, is to administer general anesthesia, and then insert fine tubes or needles into the tumor. These tubes or needles contain heating elements and radioactive sources.

Each treatment approach has advantages and disadvantages. Certain types of cancer are more appropriately treated by one method than another. Examples are provided in the table on Page 56.

Type of Hyperthermia	Appropriate Cancers for Hyperthermia Treatment
Local, superficial	Breast cancer recurrent in the chest wall
	Cancer in superficial lymph nodes such as neck, armpit, or groin
	Skin cancers, especially melanomas
Regional	Cancers in an arm or leg
	Cancers confined to the pelvis or abdominal wall
Whole body	Chronic lymphocytic leukemia or non-Hodgkin's lymphoma
	Widely spread metastatic cancers, in combination with certain chemotherapeutic drugs
Interstitial	Superficial cancers that can be implanted with several needles or tubes, such as recurrent cancers of the cervix, rectum, or prostate
	Superficial cancers in the neck, limbs, or torso
	Certain peripheral brain cancers

It would probably help every cancer patient if hyperthermia treatments could be given in combination with radiation therapy. In practice, however, only about one in twenty patients in radiation therapy clinics receive hyperthermia today. Over the next decade, as more effective heating devices and better temperature-control equipment become available, this ratio will probably increase.

Immunotherapy

Immunotherapy is an idea that has tantalized doctors for decades. The idea of a specific vaccine, some "magic bullet" that would seek out and destroy only cancer cells, is very appealing. Unfortunately, unlike most bacteria and viruses, most cancers are not recognized by our immune system as foreign or dangerous invaders.

Still, in the search for this magic bullet much has been learned about our very complex immune system. Many patients with sarcomas, certain kidney tumors, and a rare type of leukemia have benefited considerably from immunotherapy. Patients with these last two types of cancer are treated with *interferon*, a protein manufactured by white blood cells to fight viral infections. As we learn more about different types of interferon, we hope that some of them, especially if combined with other substances called *lymphokines*, will help patients with other, more common cancers.

During the last decade there has been considerable interest in the promising area of monoclonal antibodies. These antibodies are special for two reasons: First, they are manufactured and selected to combine with only one very specific foreign product. This means that they can be tailor-made to detect or seek out and destroy practically any molecular components in the body. Second, they are manufactured by special *immortalized* cells outside the body. The amount of any particular antibody available for research is limited only by the practical considerations of time and cost.

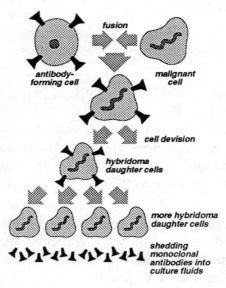

Production of monoclonal antibodies.

As shown in simplified form in the figure above, each immortalized cell comes from the fusion of one antibody-forming cell with a special type of cancer cell that gives "immortality" to the business of making just one type of antibody. Once the right cells that make the desired antibody have been produced there are several ways to use the antibody to fight cancer. For example, antibodies could be made against certain factors that cancers need to grow. Once the cancer's growth is stunted, other treatments will be more effective.

Another approach is to make antibodies against the cancer cells themselves and then to attach tiny packages of poison or radiation sources to the antibodies. While this might seem incredible, localized poisons and radiation sources have already been used successfully to destroy certain tumors in animals. Experiments with radioactive monoclonal antibodies in humans are allowing doctors to *detect* hidden cancer cells better. It is only a matter of time before they will be applied in the *treatment* of some human cancers.

CHAPTER 10

Commonly Asked Questions About Radiation Therapy

1. How do x-ray treatments work?

X-rays, given at the doses used for your daily treatments, kill dividing cells by affecting the cell's genetic material. Cancer cells are killed because they are always dividing. Certain normal (dividing) cells are destroyed as well, but some of them repair themselves after treatment.

2. Does radiation hurt?

Your x-ray treatments will not hurt, although skin irradiated for several weeks may turn pink and feel tender, as if it has been sunburned. Be very kind to your skin in the treatment area. For example, do not rub or scrub treated skin, and use only the mildest soaps with lukewarm water for washing. Wear loose, soft cotton clothing over the treated area. If bandaging is necessary, use paper tape, and try to apply it outside of the treatment area. Finally, protect this area from the sun. Cover it with light clothing or a hat before going outside, and ask your doctor about using a sunscreen lotion after your treatments have been completed.

3. Will I be radioactive?

For all practical purposes, no. A small part of your body and the treatment couch supporting you does become very weakly radioactive <u>while</u> a dose of radiation is being given. This radioactivity lasts only for a tiny fraction of a second after the linear accelerator treatment stops.

When you are given brachytherapy treatment, the radiation source remains radioactive the entire time it is in your body. During the course of brachytherapy treatment you must remain in the hospital, and visitors may be limited.

4. How often will I see my doctor during radiation treatments?

In most radiation therapy centers you will see your radiation oncologist at least once a week so that he or she can see how well you are responding to your treatments. Usually the visit will take only a few minutes. There is always a doctor available, usually your own doctor, should you wish to see him or her. In addition, your blood counts of red cells, white cells, and platelets are usually checked every 1 or 2 weeks to make sure your bone marrow has not been *suppressed* by the radiation treatments. If your bone marrow is suppressed you will be more susceptible to infection, and you may suffer *anemia*.

5. What if I have "cobalt" treatments?

Throughout this book I have based my description of teletherapy on treatment with a linear accelerator, currently the most commonly used machines for teletherapy treatment. Some radiation therapy departments, however, use cobalt machines as well as linear accelerators. Cobalt machines are very simple to operate, and they give radiation just like a linear accelerator although the radiation does not penetrate quite as deeply as the radiation from a linear accelerator. Therefore, unless your cancer is deep within the center of your chest or abdomen, the cobalt machine will work well for you.

6. What does it mean if my cancer "goes into remission"?

This simply means that your cancer is no longer "visible" by standard clinical tests (physical exam, blood tests, x-rays, etc.) There is always the possibility that a few cancer cells are still alive and able to start growing again, in spite of previous surgery or chemotherapy. The best time to treat cancers with x-rays is when they are small. Therefore, patients are often referred for *consolidative* radiation therapy to kill off any remaining cancer cells after a remission from previous therapy. Consolidative therapy builds on the partial success of the previous surgery and/or chemotherapy and completes the destruction of any cancer cells that might still be alive.

7. What should I do if I am invited to participate in a research study?

Find out as much as you can about the study from at least two people—your doctor and a skilled medical assistant, called a *data manager*, who is responsible for the accurate and safe conduct of the study. Ask them in what way the treatment plan differs from standard medical care, what the risks and benefits are, and if there will be any extra traveling, extra tests, or extra costs. If you have any doubts after speaking to your doctor and data manager, ask to take a copy of the study's consent form home with you to read. Decisions are often easier to make a day or two after you have read the consent form thoroughly.

Some people shy away from research studies because they are afraid of becoming a "guinea pig." Actually, most studies provide the highest standards of medical care. The medical community has made great advances in the treatment of leukemia and childhood cancers over the past 20 years, largely because most patients with these diseases are treated in nationally organized research studies.

Nonetheless, you are completely free to refuse to participate in a study. Your decision should have no effect on your doctor-patient relationship.

8. What special provisions are made for children who need radiation?

Everyone in the radiation therapy team tries to meet with the parents well before the treatments begin, sometimes before the child is physically examined by the radiation oncologist. The nurses and doctors discuss with the parents if it is realistic to ask the child to lie still and alone in the treatment room during radiation therapy. Restricting devices (belts or straps) are virtually never used. To keep the child motionless, comfortable styrofoam supports, favorite toys, or perhaps sedative drugs may be used. If the child is too young to understand or too frightened to cooperate, a short-acting anesthetic may be used, even daily, without seriously affecting the child. Daily practice sessions are often conducted, sometimes for two or three days, and supportive coaching from parents is often helpful.

Every effort is made to avoid irradiating the growth centers of bones in small children so later development will be as normal as possible. Of course, you must always keep the priorities in proper perspective: cancer is a potential killer, and treatments given with an aim to cure should not be compromised.

9. Will radiation therapy affect my sex life?

The answer depends on many factors, especially on the location and stage of your cancer. Radiation therapy will generally allow better organ preservation and function than surgery for cancers of the same location and stage. For example, men with early cancer of the prostate gland may be treated successfully with either radiation or surgery. One of the possible side effects of either treatment is impotence, the inability to develop an erect penis. Most evidence indicates that radiation is less likely than surgery to lead to this complication.

Some urologists claim that recent improvements in surgical techniques avoid damage to nerves needed for penile erection.

There are many other situations in which radiation might affect sexual activity. Your radiation oncologist will be glad to give you information and advice on this. You may need to remind your doctor that you are concerned about it.

10. Are radiation treatments expensive?

X-ray therapy can be costly. It requires both complex equipment and the services of highly trained health care professionals. The cost of your treatments will depend on the total number prescribed and on the complexity of the treatment plan. Most health insurance policies, including Part B of Medicare, cover charges for radiation therapy. Check with your doctor's office staff or the hospital business office to learn more about your policy and how the costs will be paid. If you need financial aid, contact the hospital social service office, the Cancer Information Service, or the local office of the American Cancer Society. They may be able to direct you to sources of help.

11. How often will I have to come back for medical checkups after x-ray therapy?

You will be informed by your doctor. The answer varies with the goals of your therapy and the risk of a recurrence or side effects. Generally, most patients are examined in follow-up visits one month after therapy is completed and at three- to six-month intervals after that for five to ten years.

CHAPTER 11

Where to Find Additional Information

Throughout this book I have recommended that you call the Cancer Information Service (CIS) 1-800-4-CANCER (800-422-6237). I cannot overemphasize the potential usefulness of this free service for discovering helpful information. Just remember that this is a national phone number, so you will first hear a recorded message giving you instructions to touch one number for answers to questions and another number to order free publications. If you have prepared a list of questions, start with the first one and take notes as you listen to each answer. If you would like to receive a publication, touch the second number. The person answering will give you guidance on which books and pamphlets are available and will take your name and address.

If you would like to read a very positive book about a patient's personal experience with cancer, ask the CIS to send you *Fighting Cancer*, written by Richard and Annette Block. Yes, Mr. Block has "been there," and he lived to write his story and offer strong advice, largely because of his own determination to seek out the best treatment available.

You may also order quite a wide selection of books from:

>Coping Magazine
>P.O. Box 1700
>Franklin, TN 37065-1700

A partial listing of books and tapes available with 1992 prices is provided below:

Siegel, Mary Ellen. *The Cancer Patient's Handbook.* McGoo's Umbrella, Saratoga, California, 1978. $14.95.

Harwell, Amy. *When Your Friend Gets Cancer.* H. Shaw, Wheaton, Illinois, 1987. $6.95.

Johnson, Judy & Klein, Linda. *I Can Cope.* DCI Publications, Minneapolis, Minnesota, 1988. $8.95

Siegel, Bernie, M.D. *Fight for Your Life* (2.5 hour videotape). $65.00. 1987.

Cox, B.G. et al. *Living with Lung Cancer.* Triad Publishing Company, 1987. $8.95.

Photopulos, Georgia & Photopulos, Bud. *Of Tears and Triumphs.* Contemporary Books, Chicago, Illinois, 1988. $16.95 hardcover, $8.95 softcover.

Bracken, Jeanne. *Children with Cancer.* Oxford University Press, Cambridge, Massachusetts, 1988. $35.00 hardcover, $10.95 softcover.

Bruning, Nancy. *Coping with Chemotherapy*. Dial Press, Gurden City, New York, 1985. $6.95.

Bradly, Jane & Nass, Susan. *Nutrition of the Cancer Patient*. Government Printing Office, Bethesda, Maryland, 1982. $15.95.

If You Love Someone Who Smokes (1 hour videotape). Films for the Humanities, Princeton, New Jersey, 1989. $39.95.

Johnson, Jacquelyn. *Intimacy: Living as a Woman after Cancer*. NC Press, Toronto, Canada, 1988. $12.95.

Love, Susan. *The Breast Book*. Addison-Wesley, Reading, Massachusetts, 1991. $21.00 hardcover, $15.00 softcover.

Chechik, Diane. *Journey To Justice*. Catalyst Publishing, Sarasota, Florida, 1987. $8.95.

CancerFax, a quick reference quide from your fax machine telephone. For up-to-the-minute cancer treatment information 24 hours a day, 7 days a week, dial 1-301-402-5874.

Radiation Therapy and You. A videotape on loan ($35.00 for 10 days) for the patient and family to view at home. Self Care Media, Inc., 935 Pleasant Drive, Ypsilanti, MI, 48197.

Other sources of information include the local chapter of the American Cancer Society, and your local library. Pamphlets are usually available through the nursing staff of your nearest cancer center or radiation oncologist's office. Finally, remember that you may seek a second or even a third opinion from most radiation, medical, and surgical oncologists. Ask the C.I.S. or PDQ for a listing of specialists in your area.

Glossary

Absorbed dose: the amount of radiation that reaches and affects the cancer or organ to be treated.

Accelerator (linear): a machine, often called a *linac*, that produces high-energy x-rays for the treatment of cancer.

Alopecia: hair loss.

Analgesia: relief of pain without loss of consciousness.

Anesthesia: a procedure involving the use of medication so that a patient is unable to feel pain, with or without loss of consciousness.

Anorexia: severe loss of appetite.

Axilla: armpit area where many lymph nodes are located that drain the arm and breast.

Benign: a type of tumor or mass with restricted growth potential; opposite of malignant.

Betatron: a machine that produces radiation used in the treatment of certain deep cancers. It is now used only rarely because of its large size and expense.

Biopsy: a procedure involving the removal of a small amount of body tissue for microscopic analysis; it is often used to make or confirm the diagnosis of cancer.

Bilateral: on both sides of the body.

Brachytherapy: a short-range type of radiation therapy. Radioactive sources are placed on or inside the cancer to be treated.

Cancer: a general term for more than 100 malignant diseases characterized by abnormal and uncontrolled growth of cells. This term includes all carcinomas and sarcomas as well as leukemias and lymphomas.

Carcinogenic: a property of some agent (for example smoke or alcohol) that could contribute to the development of cancer. (Same as "oncogenic.")

Carcinoma: the most common type of cancer. It arises from the skin or from the lining of the bowel, bronchus, or duct of a gland, such as in the breast or pancreas.

CAT scan: acronym for Computerized Axial Tomography, which provides cross-sectional x-ray images of the body, such as of the head, chest, or abdomen. It is very useful for diagnosing cancer and for planning radiation therapy treatments.

Cell: the smallest functional unit of living tissue. It may have a diameter as large as one-thousandth of an inch, but most are only one third or half that size.

Cervix: the lower part of the uterus, which projects out into the vagina.

Chemotherapy: treatment with anti-cancer drugs.

Chemotherapist: a physician who specializes in the use of drugs to treat cancer; often called a medical oncologist.

Chromosome: a strand of genes.

Clone: a collection or group of cells all derived from one cell.

Cobalt-60: a radioactive source that emits gamma rays which are similar to high energy x-rays. It is often used to treat cancer from a few feet away.

Connective tissue: the tissues of the body that bind together and support various structures of the body. Examples are bone, cartilage, muscle, and fiber.

Colon: large intestine.

Colostomy: a surgically made connection between the colon and the abdominal wall, which allows feces to bypass the rectum and the anal canal.

Colposcopy: visual examination of the vagina and cervix through a magnifying instrument that is inserted into the vagina.

CT scan: a common alternative abbreviation for CAT scan (see CAT scan).

DNA: abbreviation for deoxyribonucleic acid, the chemical name for our genes.

Dosimetrist: see Radiation Dosimetrist.

Epithelium: a layer of cells in the skin, mucous membrane, or any duct that replaces worn out cells by cell division.

Electron: a tiny particle of natural matter; it has a small negative charge.

Excision: surgical removal.

External beam radiation: a type of radiation therapy in which a machine aims radiation at a part of the patient's body; also called teletherapy.

Field: a term used in radiation oncology to describe or define a particular part of the patient's body to be treated with a x-ray beam.

Fluoride therapy: daily self-application of fluoride gel in a custom-fitted mouthpiece; this may prevent excessive tooth decay after irradiation to the head and neck region.

Frozen section: rapid technique for microscopic analysis of biopsied tissue.

Gamma rays: radiation emitted by some radioactive sources such as Cobalt-60. Gamma rays have properties identical to x-rays.

Gray (Gy): an energy unit used to measure the radiation dose to a tumor. Patients typically receive two to three grays during each radiation treatment.

Gynecology: the study of diseases of the female reproductive organs.

Hyperthermia: the elevation of tissue temperature; a cancer treatment known to enhance the curative effects of irradiation and chemotherapy. (See Chapter 9 for details.)

Hypopharynx: part of the lower throat beside and behind the larynx (or voice box).

Immunity: the condition of being resistant to a particular disease such as an infectious disease or cancer.

Implant: another name for brachytherapy.

Interferon: a protein made by certain cells, usually in response to a viral infection, that occasionally has anti-cancer as well as anti-viral properties. (See Chapter 9 for details.)

Interstitial implant: the placement of fine tubes in a gridlike pattern through tissues that contain a cancer; these tubes are filled later with radioactive sources for brachytherapy.

Intracavitary implant: the placement of a small tube within a body cavity, such as the bronchus or vagina; this tube is later filled with radioactive sources for brachytherapy.

Ionizing radiation: a type of radiation used in cancer treatment that damages the cell DNA and stops cell growth. Examples include x-rays, gamma rays, electrons, and neutrons.

Ipsilateral: on the same side of the body (opposite of contralateral).

Larynx: part of the throat used for speaking; often called the "voice box" or "Adam's apple."

Lay person: one not trained in professional (medical) matters.

Leukemia: a malignant cancer of the blood-forming tissues (bone marrow or lymph nodes), generally characterized by an over-production of white blood cells.

Linear accelerator: a machine that produces high-energy x-rays to treat cancers.

Lymph node: a collection of lymphocytes within a capsule and connected to other lymph nodes by fine lymphatic vessels; a common site for certain cancer cells to grow after travelling along lymphatic vessels.

Lymphatic system: a network of fine lymphatic vessels that collect tissue fluids from all over the body and returns these fluids to the blood. Accumulations of lymphocytes, called lymph nodes, are situated along the course of lymphatic vessels.

Lymphocyte: a type of white blood cell involved in immunity. Lymphocytes defend the body against bacteria and viruses.

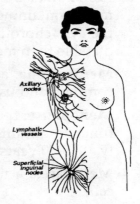

Lymphatic system

Lymphoma: a special type of cancer that begins in an altered lymphocyte. There are two broad categories of lymphomas, Hodgkin's disease and non-Hodgkin's lymphoma. Both respond favorably to radiation and chemotherapy drugs.

Malignant: opposite of benign. A property of cancers making them potentially lethal; they tend to grow without control, and they can move from one spot to another.

Mammogram: a x-ray of the breast used to detect cancer, sometimes before it can be detected by palpation. Women over 50 years old are advised to have a mammogram every year; women in their 40s every two years.

Medical oncologist: a physician with specialized training in the diagnosis and treatment of cancer with a variety of drugs and hormones. Sometimes called a chemotherapist.

Medical physicist: a physicist with specialized training for determining how to use radiation as safely as possible. Sometimes called a radiation physicist.

Melanoma: a type of cancer that begins in the pigment containing cells of a skin mole or the lining of the eye (choroid). Usually dark in color, it is less predictable than most cancers.

Meningioma: a type of brain tumor that is relatively common and usually benign.

Menopause: the time in a woman's life when ovarian function diminishes and menstrual cycles stop, usually at 45-50 years of age.

Metastases: a name for metastatic deposits of cancer cells.

Metastatic cancer: an advanced stage of cancer in which cells from the original (primary) site have spread (metastasized) to other organs.

Metastasize: to break away from the primary site and spread via the blood stream or lymphatic vessels to other sites.

MRI: an acronym for Magnetic Resonance Imaging. An MRI scan provides cross-sectional images of the body similar to CAT scan images. MRI images often give more detailed information than CAT scans.

Nasopharynx: part of the breathing passage behind the nasal cavity (see diagram).

Neurology: the study of the nervous system and its diseases.

Oncogenic: a factor or agent with the potential to cause cancer (same as carcinogenic).

Oncology: the study of cancer.

Oncologist: a doctor with specialized training in the treatment of cancer; some are specialized surgeons, some are radiation oncologists and some are medical oncologists who treat cancer with drugs (chemotherapy).

Palpate: to examine by carefully feeling with the fingers.

Pathology: the study of diseased tissues, both by gross and by microscopic examination, of tissues removed during surgery or postmortem. The pathologist is a physician with specialized training in performing and interpreting these examinations.

Nasopharynx.

Pharynx: medical term for the throat from the nasal and oral cavities above to the larynx and esophagus below (see diagram on page 77).

Platelet: a small, subcellular component of blood that helps to form blood clots.

Postmortem: Latin for "after death;" thus any postmortem examination is one that is performed on a corpse.

Primary cancer: the site where the cancer originated.

Prognosis: a qualitative (usually defined as good or poor) estimate of the future quality and/or length of life for a patient.

Prostate: a gland at the base of the bladder in males for the production of fluids that enhance the possibility of reproduction; cancer of this gland is common in elderly men (see diagram).

Protocol: a well defined treatment plan, sometimes experimental, designed to collect detailed response information about a particular group of cancer patients. It may ultimately lead to better therapy.

Proton: a positively charged particle found in every atomic nucleus.

Rad: acronym for Radiation Absorbed Dose. A unit of energy used to measure the x-ray and gamma ray energy absorbed by the body. You absorb two or three hundreths of a rad each time you have a chest x-ray.

Radiation: a general term for any form of energy emission or passage through space or matter

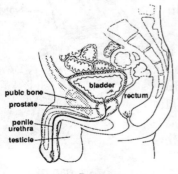
Prostate.

of invisible light, radio, TV, radar, x-rays, sound, or heat.

Radiation dosimetrist: a specialized radiation technologist who calculates the exact amount of radiation to be given to a radiation therapy patient each day. Some dosimetrists also manufacture customized blocks to shield normal tissues and/or support devices to help patients hold still during therapy.

Radiation oncologist: a physician with specialized training in the planning and delivery of radiation to treat cancer and some non-malignant conditions as well.

Radiation physicist: see medical physicist.

Radiation therapist: a potentially confusing term because it used to refer to a radiation oncologist (a physician), but now it is often used to refer to a radiation therapy technologist.

Radiation therapy: the use of various forms of radiation to treat human diseases; includes teletherapy and brachytherapy.

Radiation therapy nurse: a nurse with extra training in the support and care of patients receiving radiation therapy.

Radiation therapy technologist: a technician trained to accurately deliver the doses of radiation prescribed by the radiation oncologist, under the supervision of radiation physicists and radiation oncologists. Also known as a radiation therapist.

Radioactive: a property of all unstable elements which regularly change (or "decay") to an altered state by releasing energy in the form of photons (gamma rays) or particles (electrons, alpha particles, etc.).

Radiologist: a physician with specialized training in the "reading" of diagnostic x-rays and other techniques for imaging the body.

Recurrence: the return of a cancer after all detectable traces had been removed by primary therapy; recurrences may be local (near the primary site) or distant (metastatic).

Resection: surgical removal; in relation to cancer resection, the pathologist often indicates if the outer margins of the resection had no cancer cells present or was "negative."

Sarcoma: a type of cancer derived from connective bone or fat tissues. Examples include malignant fibrous histiocytoma, fibrosarcoma, osteogenic sarcoma, liposarcoma, etc.

Scan: a diagnostic test usually involving the movement or scanning of a detector to produce a picture. Examples include ultrasound, nuclear medicine scan, computer-assisted tomography (CAT), or magnetic resonance imaging (MRI) scans.

Secondary cancer: arising from a primary cancer; metastatic cancer.

Self-hypnosis: a passive condition in which there is increased responsiveness to suggestions and commands made by yourself. Also called hypnosis by autosuggestion, this can improve a cancer patient's mental outlook and tolerance of therapy.

Side effects: reactions to treatment caused by its effect on normal body cells; they are usually temporary and reversible, but some late side effects are permanent. Cells that divide most frequently (e.g., those that line hair follicles, oral mucous membranes, and bowel mucous membranes) are the ones most likely to be affected.

Simulation: part of the treatment planning process during which special x-rays are taken of the treatment area, and skin marks are made for guiding subsequent x-ray therapy. See pages 34-37 for details.

Teletherapy: treatment in which the radiation source is some distance (several feet) from the body (as opposed to brachytherapy): linear accelerators, cobalt machines and betatrons are examples of machines that provide teletherapy.

Tumor: an abnormal tissue mass in the body. Many masses are benign, but malignant masses are cancers. Doctors often use the word "tumor" instead of "cancer" to reduce patient anxiety. If this confuses you, ask for clarification.

X-ray: a type of ionizing radiation produced by any x-ray machine, linear accelerator or betatron. Also a common name for the images made with low-energy x-rays used in diagnostic radiology. (Some experts call them "Roentgenograms.") High energy x-rays are used for most radiation treatments in order to penetrate more deeply into the body and to deliver a uniform dose throughout the diseased tissue.

Appendix A
Comprehensive Cancer Centers

Alabama
Cancer Center
University of Alabama
University Station
1824 Sixth Avenue South
Birmingham, AL 35294
205-934-5077

Arizona
Arizona Cancer Center
University of Arizona
College of Medicine
1501 N. Campbell Avenue
Tucson, AZ 85724
602-626-7925

California
Kenneth Norris Jr. Comprehensive Cancer Center
Univ. of Southern California
P.O. Box 33800
Los Angeles, CA 90033-0800
213-224-6465

Jonsson Comprehensive
Cancer Center
UCLA
10833 LeConte Avenue
Los Angeles, CA 90024-1781
213-825-5268

Connecticut
Yale Comprehensive Cancer Center
School of Medicine
333 Cedar Street
New Haven, CT 06510-8028
203-785-4095

District of Columbia
Lombardi Cancer
Research Center
Georgetown University
Medical Center
3800 Reservoir Road N.W.
Washington, DC 20007
202-687-2110

Florida
Sylvester Comprehensive
Cancer Center
University of Miami
Medical School
P.O. Box 016960
Miami, FL 33101
305-548-4810

Maryland
Johns Hopkins
Oncology Center
600 N. Wolfe Street
Baltimore, MD 21205
410-955-8822

Massachusetts
Dana-Farber Cancer Institute
44 Binney Street
Boston, MA 02115
617-732-3636

Michigan
Meyer L. Prentis
Comprehensive Cancer Center
Wayne State University
3990 John R Street
Detroit, MI 48201
313-745-8870

University of Michigan
Cancer Center
101 Simpson Drive
Ann Arbor, MI 48109-0752
313-936-2516

Minnesota
Mayo Comprehensive
Cancer Center
Mayo Clinic
200 First Street SW
Rochester, MN 55905
507-284-4718

New Hampshire
Norris Cotton Cancer Center
Dartmouth-Hitchcock
Medical Center
2 Maynard Street
Hanover, NH 03756
603-646-5505

New York
Memorial Sloan-Kettering
Cancer Center
1275 York Avenue
New York, NY 10021
212-639-6561

Roswell Park Cancer Institute
Elm & Carlton Streets
Buffalo, NY 14263
716-845-5770

Columbia University
Comprehensive Cancer Center
College of Physicians
& Surgeons
701 W. 168th Street
New York, NY 10032
212-305-6921

Cancer Center
New York University
Medical Center
550 First Avenue
New York, NY 10016
212-263-5927

North Carolina

Duke Comprehensive
Cancer Center
Jones Research Building
Research Drive
Durham, NC 27710
919-684-3377

Lineberger Comprehensive
Cancer Center
University of North Carolina
School of Medicine
at Chapel Hill
Chapel Hill, NC 27599-7295
919-966-3036

Comprehensive Cancer Center
of Wake Forest University
Bowman Gray School
of Medicine
300 S. Hawthorne Road
Winston-Salem, NC 27103
919-748-4464

Ohio

Comprehensive Cancer Center
Arthur G. James
Cancer Hospital
Ohio State University
300 West 10th Avenue
Columbus, OH 43210-1240
614-293-5485

Pennsylvania

Fox Chase Cancer Center
7710 Burholme Avenue
Philadelphia, PA 19111
215-728-2781

University of Pennsylvania
Cancer Center
3400 Spruce Street
Philadelphia, PA 19104-4283
215-662-6334

Pittsburgh Cancer Institute
University of Pittsburgh
200 Meyran Avenue
Pittsburgh, PA 15213-3305
412-647-2072

Texas

University of Texas
M.D. Anderson Cancer Center
1515 Holcombe Boulevard
Houston, TX 77030
713-792-7500

Vermont

Vermont Regional
Cancer Center
University of Vermont
1 South Prospect Street
Burlington, VT 05401-3498
802-656-4414

Washington
Fred Hutchinson
Cancer Center
1124 Columbia Street
Seattle, WA 98104
206-667-4302

Wisconsin
University of Wisconsin
Comprehensive Cancer Center
600 Highland Avenue
Madison, WI 53792
608-263-8610

Appendix B
Radiation Oncology Residency Training Hospitals

Alabama
University of Alabama
Medical Center
Department of Radiation
Oncology
619 S. 19th Street
Birmingham, AL 35233

Arizona
Uniersity of Arizona
Affiliated Hospitals
Department of Radiation
Oncology
University of Arizona Health
Science Center
Tucson, AZ 85724

California
Loma Linda University
Affiliated Hospitals
Radiation Medicine Department
Loma Linda University
Medical Center
Loma Linda, CA 92354

Kaiser Permanente
Medical Center
Radiation Oncology
4950 Sunset Boulevard
Los Angeles, CA 90027

University of Southern
California Affiliated Hospitals
USC Department of Radiation
Oncology
1441 Eastlake Avenue,
Room 019
Los Angeles, CA 90033

UCLA Affiliated Hospitals
Department of Radiation
Oncology
UCLA Center for Health
Sciences
10833 LeConte Avenue
Los Angeles, CA 90024

University of California,
Irvine
Division of Radiation
Oncology
101 City Drive
Orange, CA 92668

St. Mary's Hospital & Medical
Center Integrated Hospitals
Department of Radiation
Oncology
St. Mary's Hospital
450 Stanyan Street
San Francisco, CA 94117

University of California, San Francisco
Department of Radiation Oncology L-75
University of California Hospital,
3rd and Parnassus Streets
San Francisco, CA 94143

Stanford University Affiliated Hospitals
Department of Radiation Oncology
Stanford University Medical Center
300 Pasteur Drive
Stanford, CA 94305

Connecticut
Yale-New Haven Hospital
Department of Radiation Therapy
Yale-New Haven Hospital
20 York Street
New Haven, CT 06504

District of Columbia
George Washington University Affiliated Hospitals
Division of Radiation Oncology & Biophysics,
George Washington Univesity Medical Center
901 23rd Street
Washington, DC 20037

Georgetown University Hospital
Department of Radiation Medicine
Georgetown University Medical Center
3800 Reservoir Road, N.W.,
Washington, DC 20007

Howard University Hospital
Department of Radiotherapy
Howard University Hospital
2041 George Avenue
Washington, DC 20060

Florida
University of Florida College of Medicine at Shands Hospital
Department of Radiation Oncology
BoxJ-385, J. Hillis Health Center, Shands Hospital,
Gainesville, FL 32610

University of Miami Affiliated Hospitals
Department of Radiation Oncology
Jackson Memorial Hospital North Wing 1
1611 Northwest 12th Avenue
Miami, FL 33136

Georgia
Medical College of Georgia Hospitals
Georgia Radiation Therapy Center
Building HK
Augusta, GA 30912-3965

Illinois

Northwestern University Medical School Affiliated Hospitals
Department of Radiation Oncology
Northwestern Memorial Hospital
250 E. Superior Street
Chicago, IL 60611

Rush-Presbyterian-St. Luke's Medical Center
Department of Ther. Radiology
1753 W. Congress Parkway
Chicago, IL 60612

University of Illinois Affiliated Hospitals
Radiology-Radiation Oncology (M/C 931),
P.O. Box 6998
Chicago, IL 60680

Michael Reese/University of Chicago Center for Radiation Therapy
Department of Radiation Oncology
University of Chicago Hospitals and Clinics
5841 S. Maryland, Box 442
Chicago, IL 60637

Loyola University Medical Center
Department of Radiotherapy
2160 S. First Avenue
Maywood, IL 60153

Indiana

Department of Radiation Oncology
Indiana University School of Medicine
535 Barnhill Drive
Indianapolis, IN 46202-5289

Iowa

University of Iowa Hospitals and Clinics
Division of Radiation Oncology
Iowa City, IA 52242

Kansas

University of Kansas Medical Center
Department of Radiation Oncology
39th and Rainbow Boulevard
Kansas City, KS 66103

Kentucky

University of Kentucky Medical Center
Department of Radiation Medicine
800 Rose Street, Room C-15
Lexington, KY 40536

University of Louisville Affiliated Hospitals

Department of Radiation
Oncology
J. Graham Brown Cancer
Center
529 S. Jackson
Louisville, KY 40202

Maryland
The Johns Hopkins Hospital
Department of Radiation
Oncology
600 N. Wolfe Street
Baltimore, MD 21205

University of Maryland
Hospital
Department of Radiation
Oncology
22 S. Greene Street
Baltimore, MD 21201

Uniformed Services University Affiliated Hospitals
(N.C.I.)
Department of Radiation
Oncology
National Cancer Institute,
Bldg. 10/B3B69
9000 Rockville Pike
Bethesda, MD 20892

Massachusetts
Joint Center for Radiation
Therapy - Harvard Medical
School
50 Binney Street
Boston, MA 02114

Massachusetts General
Hospital-Harvard Medical
School
Department of Radiation
Oncology
Boston, MA 02114

Tufts University Affiliated
Hospitals
Department of Radiation
Oncology
New England Medical Center
and Tufts University School of
Medicine
750 Washington Street
Boston, MA 02111

Michigan
University of Michigan
Affiliated Hospitals
Department of Radiation
Oncology
U.H. B2C438, Box 0010
1500 E. Medical Center Drive
Ann Arbor, MI 48109

Henry Ford Hospital
Department of Radiation
Oncology
2799 W. Grand Boulevard
Detroit, MI 48202

Wayne State University
Affiliated Hospital
Department of Radiation
Oncology
Harper Grace Hospital

3990 John R. Street
Detroit, MI 48201

William Beaumont Hospital
Department of Radiation
Oncology
3601 W. Thirteen Mile Road
Royal Oak, MI 48073

Minnesota
University of Minnesota
Affiliated Hospitals
Department of Ther. Radiology-Radiation Oncology,
University of Minnesota
Hospital
Box 494 UMHC, Harvard
Street at East River Road,
Minneapolis, MN 55455

Mayo Graduate School of
Medicine
Division of Radiation
Oncology
Mayo Clinic
200 First Street, S.W.
Rochester, MN 55905

Missouri
Washington University
Affiliated Hospitals
Radiation Oncology Center
Mallinckrodt Institute of
Radiology
510 S. Kingshighway
St. Louis, MO 63110

New Jersey
UMDNJ-Robert Wood
Johnson Medical School
Affiliated Hospitals
Department of Radiation
Oncology
Cooper Hospital University
Medical Center
One Cooper Plaza
Camden, NJ 08103

St. Barnabas Medical Center
Department of Radiation
Oncology
94 Old Short HIlls Road
Livingston, NJ 07039

New York
Beth Israel Medical Center
Department of Radiation
Oncology
1st Avenue and 16th Street
New York, NY 10003

Memorial Sloan-Kettering
Cancer Center
Department of Radiation
Oncology
1275 York Avenue
New York, NY 10021

New York University
Medical Center
Division of Radiation
Oncology
Tisch Hospital
560 First Avenue
New York, NY 10016

Columbia-Presbyterian
Medical Center
Department of Radiation
Oncology
Presbyterian Hospital
622 W. 168th Street
New York, NY 10032

Albert Einstein College of
Medicine Affiliated Hospitals
Department of Radiation
Oncology
Montefiore Medical Center
111 E. 210th Street
Bronx, NY 10467

Methodist Hospital of
Brooklyn
Department of Radiation
Therapy
506 6th Street
Brooklyn, NY 11215

SUNY Health Science Center
at Brooklyn
Department of Radiation
Oncology
450 Clarkson Avenue
Brooklyn, NY 11203

Mount Sinai Medical Center
Department of Radiation
Oncology
One Gustave L. Levy Place,
Box 1236
New York, NY 10029

New York Medical College
Department of Radiation
Medicine
Westchester County
Medical Center
Macy Pavilion West
Valhalla, NY 10595

University of Rochester
Cancer Center and Affiliated
Hospital
Department of Radiation
Oncology
Strong Memorial Hospital
601 Elmwood Avenue,
Box 647
Rochester, NY 14642

SUNY Health Science Center
at Syracuse
Radiation Oncology Division
750 E. Adams Street
Syracuse, NY 13210

North Carolina

University of North Carolina
Hospitals
Department of Radiation
Oncology
110 Manning Drive
Chapel Hill, NC 27514

Duke University Affiliated
Hospitals
Department of Radiation
Oncology
Duke University Medical
Center, Box 3085
Durham, NC 27710

Bowman Gray School of
Medicine Affilated Hospitals
Department of Radiation
Therapy
Bowman Gray School of
Medicine
300 S. Hawthorne Road
Winston-Salem, NC 27103

Ohio
University of Cincinnati
Hospital Group
Division of Radiation
Oncology
University of Cincinnati
Hospital
234 Goodman Street, Mail
Location 757
Cincinnati, OH 45267-0757

Case Western Reserve
University Hospitals
Division of Radiation
Oncology
University Hospitals of
Cleveland
2078 Abington Road
Cleveland, OH 44016

Cleveland Clinic Foundation
Department of Radiation
Therapy
9500 Euclid Avenue
Cleveland OH 44106

Ohio State University
Hospitals
Division of Radiation
Oncology

The Arthur G. James Cancer
Hospital and Research
Institute
300 W. Tenth Avenue
Columbus, OH 43210

Oklahoma
University of Oklahoma
Health Science Center
Department of Radiation
Therapy
Oklahoma Teaching
Hospitals, O.M.H.
P.O. Box 26307
Oklahoma City, OK 73126

Oregon
Oregon Health Sciences
University Affiliated
Hospitals
Department of Radiation
Oncology
3181 S.W. Sam Jackson
Park Road
Portland, OR 97201

Pennsylvania
Albert Einstein Medical
Center - Northern Division
Department of Radiation
Therapy
5501 Old York Road
Philadelphia, PA 19141

Hahnemann University
Hospital
Department of Radiation

Oncology and Nuclear Medicine
230 N. Broad Street
Philadelphia, PA 19102

Thomas Jefferson University Hospital
Department of Radiation Oncology and Nuclear Medicine
11th and Walnut Streets
Philadelphia, PA 19107

University of Pennsylvania Affiliated Hospital
Department of Radiation Oncology
University of Pennsylvania Hospital
3400 Spruce Street
Philadelphia, PA 19104

Allegheny General Hospital
Department of Ther. Radiology
320 E. North Street
Pittsburgh, PA 15212

Puerto Rico
University of Puerto Rico Affiliated Hospital
Radiation Oncology Division
University of Puerto Rico Cancer Center, Medical Sciences Campus,
P.O. Box 5067
San Juan, PR 00936

South Carolina
Medical University of South Carolina
Department of Radiation Oncology
171 Ashley Avenue
Charleston, SC 29425

Texas
University of Texas Medical Branch Hospitals
Department of Radiation Therapy
McCullough Building, Room 197, Rt. G11
Galveston, TX 77550

Baylor College of Medicine Affiliated Hospitals
Section of Radiation Therapy
One Baylor Plaza
Houston, TX 77030

University of Texas M.D. Anderson Cancer Center
Department of Radiotherapy
M.D. Anderson Hospital
Box 97, 1515 Holcombe Blvd.
Houston, TX 77030

University of Texas Health Science Center at San Antonio
Division of Radiation Oncology
7703 Floyd Curl Drive
San Antonio, TX 78284

Utah

LDS Hospital
Radiation Center
8th Avenue & C Street
Salt Lake City, UT 84143

University of Utah Medical Center
Division of Radiation Oncology
50 North Medical Drive
Salt Lake City, UT 84132

Virginia

University of Virginia
Division of Radiation Oncology
University of Virginia Hospital
P.O. Box 383
Charlottesville, VA 22908

Eastern Virginia Graduate School of Medicine
Department of Radiation Oncology
Norfolk General Hospital
600 Gresham Drive
Norfolk, VA 23507

Virginia Commonwealth University MCV Affiliated Hospitals
Department of Radiation Therapy and Oncology
Box 58
Medical College of Virginia Hospital
Richmond, VA 23298

Washington

University of Washington Affiliated Hospitals
Department of Radiation Oncology
University of Washington Hospital
1959 N.E. Pacific Street
Seattle, WA 98195

Wisconsin

University of Wisconsin Hospitals and Clinics of Madison
Section of Radiation Oncology
University of Wisconsin Hospital
600 Highland Avenue
K4/B100
Madison, WI 53792

Medical College of Wisconsin
Radiation Oncology Department
Milwaukee County Medical Complex
8700 W. Wisconsin Avenue
Milwaukee, WI 53226

Index

Adjuvant chemotherapy 26
Anemia 61
Antibodies 57-59
Benign 1-2
Biopsy 14
Bone scan 7, 28
Brachytherapy 50, 51, 52, 61, 72, 73, 75, 80, 82
Cancer Information Service (CIS) 13, 67
Carbolfuchsin 34
Carcinoma 1, 4
CAT Scan 28, 31, 32, 72, 73, 77
Check Films 41
Chemotherapy 14, 20, 21, 25-27, 48, 54, 62
Children (treatment of) 16, 48, 49, 63-64
Cobalt 62, 73, 74, 82
Consolidative radiation therapy 62
Cure 1, 7, 8, 19, 20, 21, 24, 25, 29, 42, 48, 51, 52, 64
Cure Rate 8
Data manager 63
Daughter cells 2
Decay 74, 80
Denial 5
Depression 5
Detection of cancer 3, 21, 23, 26, 47, 58, 59, 76, 80, 81
Developing a plan 6-7
Diarrhea 26, 45, 46

Dosimetrist 31, 41, 44, 73, 79
Dry mouth 47
Fatigue 46
Fluoroscope 36
Foreign cells 20, 57, 58
Gantry (of a linear accelerator) 40
Grief 8-9
Guilt 5-6
Hair loss 26, 46
Hope 6
Hormonotherapy 27
Hyperthermia 20, 27, 54-57, 74
I Can Cope 10
Immortalized cell 58
Immunotherapy 20, 27, 54, 57-59
Interferon 57, 75
Interstitial Hyperthermia 55
Intraoperative radiation therapy 52-53
Intravenous chemotherapy 25
Larynx 23, 75
Lasers 40
Leukemia 1, 4, 56, 57, 63
Linear accelerator (linac) 38, 71
Local Hyperthermia 54
Lymphatic System 4, 76
Lymphokines 57
Lymphoma 1, 4, 23, 45, 56
Malignant 1-2
Mammography 3
Medical Oncologist 14, 26
Medical Physicist 31, 33

Metastases 4, 24, 77
MRI Scan 28, 77
Nausea 26, 46-47
Oncogenic 72, 78
One Day at a Time 10
Palliate (palliation) 7
Palliative therapy 20
Radiation Oncologist 14, 23, 28, 29, 30, 32, 36, 39, 41, 42, 44, 45, 46, 47, 51, 61, 64, 65, 70, 78, 80
Radiation Therapist 32, 80
Radical operation 21
Regional hyperthermia 55
Remission 62
Resentment 5
Resident 17
Sarcoma 1, 4, 57
Scans 7, 15, 28, 47, 81
Scar tissue 47
Self image 9
Side effects 7, 22, 24, 25, 26, 27, 29, 30, 42, 45-50, 64, 65
Simulation 31, 34-37, 39, 44, 81
Staging 7
Stanford study 10-11
Stereotactic therapy 51-52
Stool test kit 3
Studies (research) 13, 18, 63
Suppression of the bone marrow 61
Surgery 14, 20, 21-22, 23, 27, 42, 48, 62
Teletherapy 50, 52, 53, 62, 73, 74, 80, 82
Testing 14-15
Treatment centers 17-18
Treatment Field 36

Treatment planning 6, 14, 29, 30, 33, 35, 41, 42, 46, 51, 52, 63, 65
Tumor 1, 2, 21, 23, 24, 27, 31, 37, 48, 50, 51, 52, 54, 55, 57, 59, 74, 77, 82
Warning signals (of cancer) 2